# POWER UP WITH ENERGY!

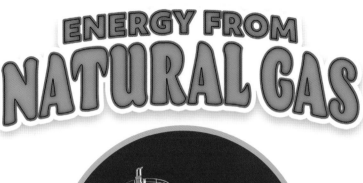
# ENERGY FROM NATURAL GAS

by Karen Latchana Kenney

Consultant: Beth Gambro
Reading Specialist, Yorkville, Illinois

BEARPORT
PUBLISHING

Minneapolis, Minnesota

# Teaching Tips

## Before Reading

- Look at the cover of the book. Discuss the picture and the title.

- Ask readers to brainstorm a list of what they already know about natural gas and energy. What can they expect to see in this book?

- Go on a picture walk, looking through the pictures to discuss vocabulary and make predictions about the text.

## During Reading

- Read for purpose. Encourage readers to think about gas and energy and the roles they play in our daily lives as they are reading.

- Ask readers to look for the details of the book. What are they learning about natural gas?

- If readers encounter an unknown word, ask them to look at the sounds in the word. Then, ask them to look at the rest of the page. Are there any clues to help them understand?

## After Reading

- Encourage readers to pick a buddy and reread the book together.

- Ask readers to name one reason to use natural gas and one reason to not use natural gas. Go back and find the pages that tell about these things.

- Ask readers to write or draw something they learned about energy from natural gas.

**Credits:**

Cover and title page, © HAYKIRDI/iStock; 3, © cigdem/Shutterstock; 5, © Ulga/iStock; 6–7, © vovashevchuk/iStock; 10–11, © Natalia Darmoroz/iStock; 13, © Maximov Denis/Shutterstock; 15, © Monkey Business Images/Shutterstock; 17, © ranplett/iStock; 18–19, © rmitsch/iStock; 20–21, © HAYKIRDI/iStock; 22, © WPAINTER-Std/Shutterstock; 23BL, © PHARAON/iStock; 23BR, © borchee/iStock; 23TL, © MarianVejcik/iStock; 23TR, © Joey Ingelhart/iStock

*Library of Congress Cataloging-in-Publication Data*

Names: Kenney, Karen Latchana, author.
Title: Energy from natural gas / Karen Latchana Kenney.
Description: Minneapolis, Minnesota : Bearport Publishing Company, [2022] |
Series: Power up with energy! | "Bearcub books." | Includes bibliographical references and index.
Identifiers: LCCN 2020054976 (print) | LCCN 2020054977 (ebook) | ISBN
9781647478667 (library binding) | ISBN 9781647478735 (paperback) | ISBN
9781647478803 (ebook)
Subjects: LCSH: Gas as fuel--Juvenile literature. | Natural gas--Juvenile
literature.
Classification: LCC TP350 .K46 2022 (print) | LCC TP350 (ebook) | DDC
665.7--dc23
LC record available at https://lccn.loc.gov/2020054976
LC ebook record available at https://lccn.loc.gov/2020054977

For more information, write to Bearport Publishing, 5357 Penn Avenue South, Minneapolis, MN 55419.
Printed in the United States of America.

# Contents

**Cooking Food** . . . . . . . . . . . . . . . . . . . 4

Energy from Natural Gas . . . . . . . . . . . . . . . . . . 22

Glossary . . . . . . . . . . . . . . . . . . . . . . . . . . . . . . 23

Index . . . . . . . . . . . . . . . . . . . . . . . . . . . . . . . . 24

Read More . . . . . . . . . . . . . . . . . . . . . . . . . . . 24

Learn More Online . . . . . . . . . . . . . . . . . . . . . 24

About the Author . . . . . . . . . . . . . . . . . . . . . . 24

# Cooking Food

Let's make pasta!

First, we need to heat water.

How do we do that?

We can use **natural** gas!

5

We need **energy**.

Energy gives things power.

Energy makes our stove work.

We can get energy when we **burn** gas.

Gas is like air.

You cannot see natural gas.

You cannot feel it.

It is found deep in the ground.

Gas comes from plants and animals that lived long ago.

They died and slowly changed into gas.

The gas now fills spaces between rocks.

To get to the gas, a big **drill** digs a hole.

Gas moves up the hole.

We take the gas when it gets to the top.

A drill

We burn gas to power machines.

It can help us cook our food.

We can use it as we heat our homes.

We use natural gas because it is easy to find.

And it is easy to send to homes.

It does not cost a lot to use.

There are bad things about gas, too.

One day we will run out of natural gas.

Gas can make our air dirty.

20

People use natural gas every day.

But one day there will not be any left.

Then, we will have to use other kinds of energy.

# Energy from Natural Gas

Follow along as natural gas is made.

**1**
Long ago, plants and animals died in the ocean.

**2**
Dirt covered the plants and animals. It pushed down.

**3**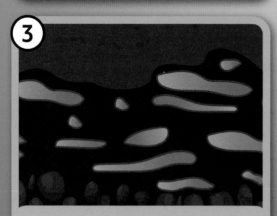
The dead things got hot.

**4**
Millions of years went by. Some of the dead things turned into natural gas.

# Glossary

**burn** to use as fuel, such as for a fire

**drill** a large machine that digs deep into the ground

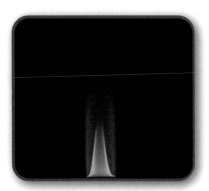

**energy** power that makes things work

**natural** made by nature not people

# Index

**burn**  6, 14

**drill**  12–13

**energy**  6, 21–22

**power**  6, 14

**rocks**  10

**stove**  6

# Read More

**O'Brien, Cynthia.** *Energy Everywhere (Full STEAM Ahead!)*. New York: Crabtree, 2020.

**Olson, Elsie.** *Natural Gas Energy (Earth's Energy Resources)*. Minneapolis: Abdo Publishing, 2019.

# Learn More Online

1. Go to **www.factsurfer.com**
2. Enter "**Gas Energy**" into the search box.
3. Click on the cover of this book to see a list of websites.

# About the Author

Karen Latchana Kenney likes biking and reading. She tries to find ways to use less energy every day.